GREAT WESTERN REVIVAL

Western Locomotives in the Preservation Era

John Maybery

AMBERLEY

First published 2015

Amberley Publishing
The Hill, Stroud
Gloucestershire, GL5 4EP

www.amberley-books.com

ISBN 978 1 4456 3987 1 (print)
ISBN 978 1 4456 3992 5 (ebook)

British Library Cataloguing in Publication Data.
A catalogue record for this book is available from the British Library.

Typesetting by Amberley Publishing.
Printed in the UK.

Contents

Introduction

My trainspotting days started in 1946/7 – the last years of the Great Western Railway. Living in Bristol, the GWR was my railway of first choice and most Saturday mornings I could be found at the west end of No. 4 platform at Bristol Temple Meads. For the price of a 1*d* platform ticket and avoiding the Railway Police you could have a cheap morning out. In theory a ticket only lasted for one hour, but one could always go out of the station and renew it immediately. With a small group of like-minded friends I ranged far and wide in the West Country and South Wales in a quest for new 'cops'. I still have my ABCs from those days – all neatly underlined and proving invaluable in complying this book.

By the early 1950s other interests were beginning to take priority – work, further education, girlfriends and impending National Service. As the decade moved on I married and was busy establishing myself in a new job. My new employment took us to Frome in Somerset, which despite the Beeching cuts still retained a railway station with an overall station roof designed by Brunel; I am pleased to say that this is now a listed building. By this time my interest in railways was on the 'back burner'; however, in the early 1970s I was driving on the A361 through the village of Cranmore and spotted a large building which resembled a Great Western engine shed. This was David Shepherd's embryonic East Somerset Railway. In Frome we lived not far from the West of England main line, and one Sunday morning I awoke to the sound of steam trains (or was I dreaming?) – this was the sound of David Shepherd's two locomotives, 9F 2-10-0 *Black Prince* and 4MT 4-6-0 *Green Knight*, en route to Cranmore. My interest in steam was reborn.

Accompanied by my two small sons and an understanding wife, I started to relive my boyhood interest in steam. I also linked up with my trainspotting friends of the 1940s/50s and we made many trips around the country to preserved lines. I was pleasantly surprised to find Great Western locomotives in many locations – not just confined to the West Country. I have taken many photographs of Great Western engines in these locations; they form the basis of this book. I am also amazed at the large amounts of money that the preserved lines are able to attract through the generosity of steam enthusiasts as many of them would never have seen these engines in revenue earning service – e.g. the cost of repairing the land slips on the Severn Valley and Gloucestershire Warwickshire railways.

Many new builds are being undertaken by enthusiasts at various steam centres and I look forward in particular to seeing a Grange, Saint, County and 47XX 2-8-0 in steam once again. I am sure the Grange will obtain a main line certificate, but the others may be confined to the preserved lines. The sight of a 47XX with ten or twelve coaches coming down the West of England main line would be a glorious sight.

I do feel, however, that there is a need for some smaller locomotives on the lines of the new build WSR Mogul No. 9351. I am sure many GW-themed lines could find room for a Bulldog or a Dean Goods.

This book is not intended to cover all the preserved Great Western engines – for example, the static exhibits at the Steam Museum at Swindon are not included. I have however included in many cases several shots of the same locomotive from different angles and in different locations.

Since my early trainspotting days I have always taken photographs, and over the years I have updated my cameras from the primitive equipment of the early days. All the photographs in this book were taken with Canon SLR cameras, the AV1 and AE1P models – alas now obsolete in this digital age. The photographs cover the period 1982–2009.

Finally, my heartfelt thanks to my wife Joyce, who has painstakingly typed the manuscript for this book. By the time we had finished she could at least tell the difference between a pannier tank and a saddle tank!

I hope you enjoy my book.

John Maybery
September 2015

Great Western 150 Celebrations

1985 saw the 150th anniversary of the Great Western Railway. British Rail and several steam centres held special events to celebrate. I visited events held at Westbury, Didcot Steam Centre and Taunton station – copies of the admission tickets are shown below. At the Westbury diesel depot David Shepherd's 9F *Black Prince* was on display; I imagine it had just steamed over from Cranmore for a couple of days – can you imagine the red tape involved if this happened today? Incidentally, there was a shuttle bus service from Westbury to Cranmore at the cost of £1.50!

150th ANNIVERSARY of the
GREAT WESTERN RAILWAY

№ 5362

WESTBURY
OPEN
DAY

SUNDAY
MAY 5th
1985

SOUVENIR ADMISSION TICKET

Admission Price £1.50 Adult, 50p Child

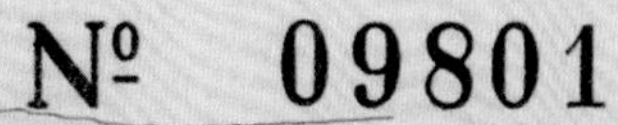

Great Western Steam 150
Didcot Railway Centre 18 May-2 June 1985

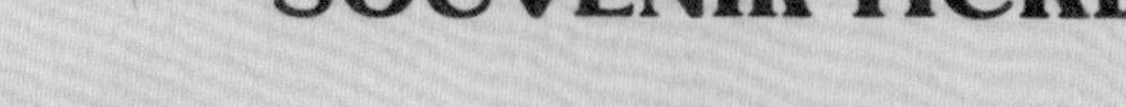

SOUVENIR TICKET

ADMIT ONE
VALID FOR ENTRY ON ANY ONE DAY BETWEEN 18 MAY AND 2 JUNE 1985

193 **TAUNTON**

Souvenir Platform Ticket

7th JULY, 1985

Admit One Person to view GWR Steam Locomotives:

Drysllwyn Castle *and* **Hagley Hall**

Above: Westbury depot, 5 May 1985. The original 4-2-2 *Iron Duke* was built in 1847 and withdrawn in 1871. The replica shown in this photo was built by Resco Railways as part of the GW 150 celebrations, being built from two Hunslet Austerity locos, and is part of the NRM collection. *Iron Duke* is on a Flatrol wagon with its tender on a second wagon coupled behind. It was due to return to Didcot in 2013 for cosmetic restoration.

Below: Another view of *Iron Duke*, showing the wood-covered boiler and fire box. The Flatrol wagon carries the designation 'to carry the *Iron Duke*'.

Above: This view shows the fire box and primitive cab area of *Iron Duke* – no protection on the footplate whatsoever – you can imagine the state of the crew after a trip from Paddington to Bristol on a windy rainy day.

Below: David Shepherd's 2-10-0 *Black Prince* has been brought over from Cranmore. Not a Great Western engine, but built at Swindon in March 1960 and withdrawn in March 1965. *Black Prince* probably just steamed over from Cranmore for the weekend – imagine the paperwork required these days.

Above: Another view of *Iron Duke* and *Black Prince*. The 2-10-0 towers over the diminutive broad gauge loco.

Below: The Western Region managed to name some of the Brush Type 4 diesels and this is the nameplate of *Isambard Kingdom Brunel*, complete with GW coat of arms. At that time the locomotive was numbered D1662.

Above: Didcot Railway Centre, 1 June 1985. Many GW locomotives were on display. This photograph shows Castle Class No. 5051, then named *Drysllwyn Castle*.

Below: Manor Class No. 7808 *Cookham Manor* stands at the approach to the turntable at the rear of Didcot shed.

Above: Mixed-traffic locomotive Churchward Mogul No. 5322 looking immaculate in unlined Brunswick green livery.

Below: Class 42XX 2-8-0T No. 5224 in black BR livery. No. 5224 stands alongside Dukedog 4-4-0 No. 3217, now carrying its original name *Earl of Berkeley*.

Above: Class 56XX 0-6-2T No. 6697 in GW livery. A powerful class of tank engine designed by C. B. Collett for working the coal traffic in South Wales, many of these new locos were replacements for the ageing locos acquired in the 1923 amalgamation.

Below: A modern Pannier tank Class 94XX No. 9466, designed by F. W. Hawksworth in 1947. Over 200 of these locomotives were built, being numbered in the 94XX, 84XX and 34XX series.

Above: A Great Western wartime build: LMS 8F 2-8-0 No. 8431, built at Swindon March 1944, used on the GW during the war and returned to the LMS in March 1947. Note the GW style numbering on the buffer beam.

Below: Foreign visitor – LNER A2 Pacific No. 532 *Blue Peter* stands on one of the demonstration lines double-heading a 51XX 2-6-2T.

Above: Taunton station, 7 July 1985: Castle Class No. 5051 and Hall Class 4930 double-head the Great Western Limited running from Bristol to Plymouth to commemorate GW 150. The Castle is in GW livery, and the Hall in BR green livery.

Below: Hall Class 4-6-0 No. 4930 *Hagley Hall* is the inside engine on the Great Western Limited.

Above: The starter signal is in the Up position and some attention is being given to the bogie of the leading coach.

Below: This photo depicts the nameplate of *Hagley Hall*, in auction terms currently worth £8,000–£10,000; at one time they could be bought directly from Swindon Works for as little as £10 per plate. I knew someone who bought two plates at this price! In the background the crew of the Castle trim the coal in her tender.

The nameplate of the immaculate *Drysllywyn Castle*. In the background is the large Taunton West End signal box. The starter signal shows that 'right away' is imminent.

Castles and Kings

Until the early to mid-1920s the principal express locomotives of the Great Western were Stars and Saints; in fact the last Star was built in early 1923. The new chief mechanical engineer, C. B. Collett, had already designed a new class of express locomotive. This was the Castle Class – in reality an enlarged Star with a bigger boiler and cylinders. The first Castle was completed in August 1923 and the last one was delivered in August 1950, a total of 171.

He then designed a further class – the Kings. These were much bigger engines than the Castles with a larger boiler, higher boiler pressure and slightly smaller driving wheels giving a tractive effort of 40,300 lbs – at that time the most powerful locomotive in the country. The Kings were built in the period 1927–1930, with a total of thirty engines.

Due to their extra weight the Kings had limited route availability, mainly working the Cornish Riviera, the Torbay Express and the Paddington–Birmingham expresses; they were not permitted through the Severn Tunnel and South Wales. The Kings were all withdrawn in 1962, being made redundant by the Warship and Western diesel hydraulics. In BR days they were all given double chimneys; in the preservation era No. 6023 has reverted to a single chimney.

The Castles had much wider route availability and were probably best known for their work on the Bristolian and the Cheltenham Spa Express. Numerous Castles have been preserved, the best known of which is *Clun Castle*.

Above: June 1990: The Great Western 'Pride of the Line' No. 6000 *King George V* is on display at Swindon Works in an exhibition organised by the railway museum. On the front of the footplate is mounted the bell presented to the engine by the Baltimore & Ohio Railway – a tour of the United States was undertaken in the late '20s.

Right: July 1985: *King George V* is at the head of the Great Western Limited at Taunton as part of the GW 150 celebrations.

Above: July 1985: Another shot of *King George V* departing Taunton at the head of the Great Western Limited.

Below: I cannot date this photograph, but it shows No. 6024 *King Edward I* on shed at the Great Western Centre, Didcot.

Above: The same unknown date with another view of No. 6024 'on shed' at Didcot. This photograph clearly shows the massive size of the Kings compared to the Castles.

Below: Date and location not known but No. 6024 is at the head of The Shakespeare Express, so the location is probably somewhere between Birmingham and Stratford-upon-Avon.

Above: June 1990: No. 6024 'on shed' at Didcot, undergoing some light repairs and in particular some attention to the safety valve casing.

Below: March 1997: No. 6024 running into Watchet station on the West Somerset Railway. The shed code on the smoke box door is 81A, indicating the Old Oak Common shed.

Above: March 1997: No. 6024 is receiving some attention at Bishops Lydeard on the WS Railway. Note the reduced height of the safety valve casing, which has been lowered to give safe clearance under bridges – it did not improve her appearance.

Below: July 1995: Two Castles on the Great Western Limited near Highbridge, Somerset. The two locomotives are No. 7029 *Clun Castle* and No. 5051 *Earl Bathhurst*.

Above: June 1998: In the early stages of restoration the boiler of No. 5029 *Nunney Castle* rests on a Flatrol wagon at Didcot Railway Centre.

Below: The date is unknown but No. 5029, now fully restored, runs round a train at the Minehead terminus of the West Somerset Railway.

Above: October 1994: No. 5029 has just pulled into Cranmore station on the East Somerset Railway. The Capitals United Express has just run down from Paddington, coming off the West of England main line at Witham Friary onto the branch for Cranmore.

Below: October 1994: Another shot of No. 5029 in the yard at Cranmore. This is the closest the locomotive will get to the real-life Nunney Castle, which is situated in the village of Nunney – about 4 miles to the east.

Above: October 1994: No. 5029 looking resplendent in Great Western livery. The Capitals Express headboard has been removed.

Below: October 1994: A final shot of *Nunney Castle* at Cranmore standing alongside David Shepherd's 9F 2-10-0 No. 92203, which at that time was based at Cranmore.

Above: June 1989: No. 5051 *Earl Bathhurst* in the yard at Didcot Railway Centre. The blue tender in the background is attached to LNER Class A4 Pacific *Sir Nigel Gresley*.

Below: Date unknown but certainly before June 1989: Another view of No. 5051 but carrying the name *Drysllwyn Castle* on the right-hand splasher. At this time the loco carried dual names – on the left-hand splasher it is titled *Earl Bathurst*.

Above: November 1989: One of the post-war Castles, No. 7027 *Thornbury Castle*, stands in the yard at Buckfastleigh on the Dart Valley Railway. I believe it was originally purchased for spares but has now been restored at Crewe Railway Centre.

Left: March 1991: Probably the best known of the preserved Castles, No. 7029 *Clun Castle*, receiving some attention in the storage sidings at Bishops Lydeard on the WSR.

1991: No. 7029 *Clun Castle* and 2-6-2T No. 4561 at the end of a working day at Bishops Lydeard on the WSR.

Mixed Traffic Locomotives

This chapter covers four classes: the Halls/Modified Halls, the Manors, the 43XX Moguls and the 2251 (Collett) classes.

The first Hall was derived from the Saint Class *St Martin*. No. 2925 was renumbered No. 4900 and given smaller driving wheels (6 feet); the remainder of the class starting with No. 4901 was built from 1928 onwards. A total of 371 engines were built, the last being No. 7929 which was built in 1950. The last seventy-one engines were built by F. W. Hawksworth and called Modified Halls, having a redesigned main frame and a plate frame bogie. The Halls were more than capable of handling express trains such as the Paddington–Weston-super-Mare Merchant Venturer.

The next class was the Manors. A total of thirty were built: twenty pre-war and the final ten post-war in 1950. Some people may query their inclusion in this chapter, but the GWR called them mixed traffic locos and BR gave them the power classification 5MT. I must say I never saw one on a freight train – they were more usually found on the mid-Wales line and as pilots on the south Devon banks.

The Mogul 2-6-0s were another prolific class of mixed traffic locos; building commenced in 1911 under G. J. Churchward and was completed in 1932 by C. B. Collett. They were all numbered in the X3XX series and the last twenty (Nos 9300–9319) were given Castle-type cabs, which much improved their appearance compared to the austere Churchward cabs. The Moguls were capable of handling any traffic and in the summer months would often be pressed into serviced to handle relief express trains. One of each type has been preserved.

The final class is the 2251 0-6-0s, usually known as the Colletts. 120 locos were built from 1930 to 1948: the first 100 were numbered Nos 2200–2299, and the final twenty numbered 3200–3219. The Colletts were equally at home on transfer goods and local passenger trains; they had a remarkable turn of acceleration with a load of two to four coaches. One example of this class has been preserved and apart from its initial home on the Severn Valley it has made many visits to other preserved lines.

Above: March 1997: Hall Class No. 4920 *Dumbleton Hall*, looking immaculate in Great Western livery, awaits her next turn of duty at Minehead station on the West Somerset Railway.

Below: May 1983: Hall Class No. 4930 *Hagley Hall* approaching Highley station on the Severn Valley Railway. *Hagley Hall* is coupled to one of the Hawksworth-designed straight-sided tenders.

Above: June 1986: Another view of No. 4930 on the Severn Valley Railway at Bewdley station. By this time she has been repainted in BR livery and the tender has reverted to the original Collett design.

Below: October 1987: Once again we see No. 4930 on the Severn Valley Railway, standing outside the shed adjacent to the Bridgnorth terminus.

Above: May 2000: Returning to the West Somerset Railway, No. 4936 *Kinlet Hall* is receiving some attention prior to taking up her next duty roster.

Below: March 1982: Hall Class No. 5900 *Hinderton Hall* is seen here at Didcot Railway Centre.

Left: March 1982: Another view of No. 5900 standing on one of the demonstration lines at Didcot Railway Centre.

Below: June 1988: Still at Didcot Railway Centre, No. 5900 *Hinderton Hall* stands outside the shed with a Western Class diesel to the rear.

Above: June 1990: The scene now moves to Llangollen, the terminus of the Llangollen Railway. Hall Class No. 5952 *Cogan Hall* stands in a bay platform in the early stages of restoration.

Below: June 1990: Hawksworth-designed Modified Hall No. 6960 *Raveningham Hall* stands at the head of a train at Bridgnorth station.

Above: September 1988: Another view of *Raveningham Hall* at Bridgnorth station, standing ready to depart with a train to Bewdley. It carries the shed code 82A – Bristol (Bath Road).

Below: June 1987: Another Modified Hall, this time No. 6998 *Burton Agnes Hall*, on the demonstration line at Didcot; behind the fence is the main Paddington–Oxford line.

Above: June 1987: Another view of *Burton Agnes Hall*, also at Didcot Railway Centre, carrying the headboard The Red Dragon, one of several named trains which ran from Paddington to South Wales.

Below: June 1989: No. 6998 *Burton Agnes Hall* at Arley on the Severn Valley Railway, crossing the Victoria Bridge.

Above: June 1991: At Didcot Railway Centre No. 6998, displaying a 'not to be moved' sign, stands alongside Standard 4-6-2 7P 3-cylinder Pacific No. 71000 *Duke of Gloucester*, the only 3-cylinder loco to be built by British Rail.

Below: August 1994: Manor Class No. 7802 *Bradley Manor* stands in the yard at Minehead station on the WSR. Unusually it is coupled to a 4,000-gallon tender, not a common practice in Great Western days.

Above: September 1984: Another Manor, this time No. 7808 *Cookham Manor*, pictured at Didcot Railway Centre. This loco is coupled to the more usual size of tender.

Below: Date not known: Manor Class No. 7808 showing a fair rate of speed on the demonstration line at Didcot Railway Centre.

Above: Once again date unknown, but No. 7808 has obviously been turned on the turntable at Didcot since the previous photograph was taken. The tender carries the Great Western 'shirt button' style logo.

Below: May 1983: Manor Class No. 7812 *Erlestoke Manor* approaching Hampton Loade on the Severn Valley Railway.

Above: May 1983: The driver of No. 7812 takes a single line token from the signalman at Highley station.

Below: May 1983: Manor Class No. 7812 at the head of a train for Bewdley about to depart Bridgnorth station.

Above: July 1985: A final shot of No. 7812 standing 'out of service' at Bridgnorth shed.

Below: June 1987: Another Manor, No. 7819 *Hinton Manor*, ready to depart Bridgnorth to Bewdley on the SVR.

Above: June 1987: A change of country – we are now at Machynlleth station where No. 7819, painted in BR mixed traffic livery, is being prepared to haul the Cardigan Bay Express to Barmouth.

Below: June 1987: At the end of its journey No. 7819 is detached from its train at Barmouth station to run round before its return journey.

Above: June 1987: Having run round its train, No. 7819 returns tender first with the headboard now attached to the rear of the tender. The leading coach appears to be an auto-car with windows at the end of the coach.

Below: March 1987: Manor class No. 7820 *Dinmore Manor* stands on the eastbound platform at Blue Anchor on the West Somerset Railway.

Above: March 1997: Manor Class No. 7820 standing in Minehead station before running onto its train for the return journey to Bishops Lydeard.

Below: March 1997: No. 7820 *Dinmore Manor* prepares to depart Bishops Lydeard with a train for Minehead.

Above: March 1997: A final shot of No. 7820 standing outside the locomotive shed at Minehead station. She is in steam so presumably is waiting to take a train back to Bishops Lydeard.

Below: June 1990: I can think of safer places to sit! This is the scene at Llangollen on the Llangollen Railway. Manor Class No. 7822 *Foxcote Manor* is running through the platform to head up the next departure.

Above: July 1993: Manor Class No. 7822 awaiting departure from Cranmore station on the East Somerset Railway.

Below: July 1993: Another view of No. 7822 at Cranmore station. Two young enthusiasts are looking onto the footplate; are they hoping to be invited on board?

Above: July 1993: At the other end of the East Somerset line there is a run-around platform called Mendip Vale. This photograph shows No. 7822 preparing to tackle a train back to Cranmore.

Below: September 1987: Manor Class No. 7827 *Lydham Manor* stands in Paignton station preparing to take a train to Kingswear.

Above: June 1999: This scene could be mistaken for Talerdigg Bank on the mid-Wales line, but it actually shows No. 7828 climbing the bank out of Washford en route to Minehead on the WSR.

Below: July 1999: The final photograph of Manor No. 7828, which has just been uncoupled from the train on the right-hand side of the picture and will run tender first back to Bishops Lydeard.

Above: June 1989: Churchward Mogul No. 5322 on one of the demonstration lines at Didcot Railway Centre; the loco is painted in unlined Brunswick green livery.

Below: June 1989: Another view of Mogul No. 5322 on the demonstration line, pulling one of the large Great Western coaches with recessed entry doors at either end of the carriage.

Above: March 1997: Collett-built Mogul No. 7325 double heads a train from Minehead into Blue Anchor station. When built, No. 7325 was originally numbered 9303; the number was changed in the early '50s.

Below: March 1987: Another view of No. 7325 standing in Blue Anchor station en route to Bishops Lydeard. The Collett-built Moguls had Castle-type cabs and originally formed a distinctive subclass numbered 9300–9309.

Above: September 1988: No. 3205 stands alongside pannier No. 6412 outside the small loco shed at Minehead station.

Below: June 1981: The 2251 Class 0-6-0s were known as the Colletts. No. 3205 is the only one of the class preserved and is seen here approaching Bewdley station on the Severn Valley Railway. It was originally purchased for preservation by Sir Gerald Nabarro, the local MP, and spent its early days on the Severn Valley Railway.

Above: September 1990: Looking immaculate in the late summer sunlight, No. 3205 stands at Minehead station having just arrived with a train from Bishops Lydeard.

Below: March 1997: At Blue Anchor station No. 3205 double-heads a Manor Class 4-6-0 with a train for Minehead. Blue Anchor is a passing point for trains and the train going in the other direction can just be seen at the left-hand side of the picture.

March 1997: Later on in the same day at Blue Anchor station, No. 3205 departs with a train for Minehead.

Freight Locomotives – Tender and Tank

The Great Western used both tank and tender engines in the freight role; I believe I am correct in saying all types have been preserved. The first tank engines were the 2-8-0Ts in the 42XX and 52XX series. Building extended from 1912 to 1940 with variations in the frames and steam pipe arrangements. Building commenced under G. J. Churchward and was completed by C. B. Collett.They were referred to as the tank version of the 2-8-0 tender engines, but they were not – the wheel base and cylinder arrangement were entirely different. As an enthusiastic railway modeller I found this attempted conversion both frustrating and expensive.

In real life some of these 2-8-0Ts were converted to 2-8-2Ts; a total of fifty-four were rebuilt by C. B. Collett in the period 1934 to 1939. They were massive engines and were prone to derailing on some of the sharper curves to be found in sidings. They were all numbered in the 72XX series – 7200 to 7253.

A smaller locomotive was the powerful 56XX 0-6-2T; in total 200 were built in the period 1924–1928. Many of them were built by sub-contractors and were numbered 5600–5699 and 6600–6699. The majority of them worked in the Welsh valleys on coal trains, and they were replacements for the many ageing locomotives taken into stock in the railway rationalisation of 1923; some, however, did have other roles such as bankers on the Westbury–Salisbury line.

The tender engines were the 2-8-0s in the 2800 and 2884 classes. They were built in the period 1903–1942. The 2884 Class were similar to the 2800 Class but with Collett-type cabs, again a modification which much improved their appearance. They were built from 1938 onwards.

Finally we come to the LMS 2-8-0 tender locomotives. They were definitely not a GWR design but were numbered 8400 to 8479 and built at Swindon as a wartime requirement. They were built in the period 1943–1945 and were all returned to the LMS by 1947, although they were all used on the Great Western during the war years.

Above: Date not known: 42XX Class 2-8-0T No. 4247 awaits her next turn of duty on the Gloucestershire Warwickshire Railway at Toddington.

Below: Another view of No. 4247 at Toddington. There were three design variants of this class of loco; this photograph illustrates the variant with external steam pipes and straight frames.

Above: July 1999: Another member of the 42XX Class, this time No. 4277, is seen here at Minehead on the West Somerset Railway.

Below: July 1999: No. 4277 has now moved to the other end of the line here at Bishops Lydeard and is in the process of topping-up her water tanks. The 42XX Class were particularly powerful engines, which is demonstrated by the D/Red Power-Weight classification above the number plate.

Above: A final shot of No. 4277 at Bishops Lydeard taking on water. This was a Churchward design of locomotive, the first of the class being built in 1910 and the final batch being built during the Second World War.

Below: August 1998: The scene has now moved to Loughborough on the Great Central Railway. This photograph depicts one of the latter-built locos, No. 5224, which is finished in unlined BR black livery.

Above: July 1998: Another photograph of No. 5224 at Loughborough. This variant illustrates the third design style with external steam pipes and raised main frames over the cylinders.

Below: September 1987: No. 5239 is seen here on the Torbay Railway at Kingswear. This railway is now of course renamed The Dartmouth Steam Railway & Riverboat Co.

Above: September 1987: A final shot of the 52XX Class. No. 5239 awaits departure from Kingswear and you will note that it has been named *Goliath*, an unknown practice in Great Western days.

Below: June 1991: 56XX Class 0-6-2T No. 5619 is shown here on the Swanage Railway in Dorset. In front of No. 5619 is the tender of preserved Southern Railway T9 No. 120.

Above: The location has now moved to Cranmore on the East Somerset Railway. 56XX Class No. 5637 has been recently restored but painted in a rather peculiar shade of green.

Below: July 1999: 56XX Class No. 5637 is ready to depart from Cranmore station on the ESR with a four-carriage train, somewhat reminiscent of the duties they undertook on the Cardiff Valley lines.

Above: July 1990: A move to the other end of the country; Class 56XX No. 6619 is seen here at Grosmont on the North Yorkshire Moors Railway.

Below: July 1990: Another view of No. 6619 emerging from the tunnel near Grosmont pulling a train of coaches in 'plum and spilt milk' livery.

Above: May 1983: Back to Cranmore and a view of No. 6634 in plain black livery. At the best you could say she has been cosmetically restored.

Below: March 1982: At the Great Western Railway Centre, Didcot, No. 6697 is 'on shed' finished in GW plain Brunswick green livery.

Above: September 1984: Another shot taken at Didcot; this time No. 6697 stands at the head of a mixed load of preserved goods wagons. In the background is preserved pannier tank No. 3738.

Below: June 1986: 28XX 2-8-0 Class No. 2857 taking on water at Bridgnorth on the Severn Valley Railway.

Above: September 1988: We have now moved forward just over two years and No. 2857 is at Bridgnorth station at the head of a rake of 'chocolate and cream' coaches awaiting departure for Bewdley.

Below: September 1988: At the end of the working day No. 2857 is 'on shed' at Bridgnorth. The loco carries the initials STJ at the front of the footplate, denoting her home shed which at some time was Severn Tunnel Junction. At this time enthusiasts were allowed to wander around at will – a practice which has now been banned, hurray for health and safety!

Above: June 1990: A final shot of No. 2857 crossing the magnificent Victoria Bridge near Arley on the Severn Valley Railway.

Below: Date unknown. This photograph shows 2884 Class 2-8-0 No. 3803 in the early stages of restoration at Bewdley on the SVR.

Above: June 1988: Seen here at Didcot is 2884 Class No. 3822. The 2884 Class was built by C. B. Collett with Castle-type cabs. In this photograph the locomotive is in wartime guise with the cab side windows blanked out.

Below: June 1991: No. 3822 is on the turntable at Didcot Railway Centre and carries the shed code CDF on the footplate, denoting that at some time during its life it was based at Cardiff Canton.

Above: June 1991: No. 3822 is on one of the demonstration lines at Didcot Railway Centre. There is a small platform at each end of the line for passengers to get on and off the train.

Below: June 1991: A final shot of No. 3822 at Didcot shed; to the right can be seen the Great Western Society's preserved diesel railcar, No. 22.

Above: September 2000: No. 2884 Class No. 3822 is shown here at Minehead station on the West Somerset Railway, carrying a 'shirt button' logo on the tender. The shed code is SPM, usually called 'spam' by train spotters but more properly St Phillips Marsh Bristol.

Below: September 2000: No. 3850, seen here at Minehead station, has just come off a train from Bishops Lydeard and is moving to the water crane in preparation for her next duty.

Above: June 1985: Great Western-built but LMS-designed 8F No. 8431 stands on the turntable at Didcot. During the war years, the 84XX series spent all their time on Great Western lines.

Left: June 1984: Now returned to LMS livery, No. 8431 is seen here on the Keighley & Worth Valley Railway. In the previous photograph dated 1985, No. 8431 was at Didcot for the GW 150 celebrations.

July 1988: Pictured at Haworth shed, No. 8431 is now being restored. As you can see, by this time she was carrying a BR number and black livery.

Prairie Tanks

The most favoured wheel arrangement for commuter, branch-line and light freight duties was the 2-6-2 Prairie tank. In GW days there were five classes, including the 3100 Class which were virtually as powerful as the Castles. Three of these classes are now preserved – they are the 5100 Class, some of which were initially built in the early 1900s and then numbered in the 3100 series. Fairly complex renumbering and rebuilding saw them finally emerge as the 5101 Class in 1929; the final build was in 1949. These locomotives were numbered in the 41XX and 51XX series.

Another class, the 6100 Class, was also built in 1929. Visually they were identical to the 5100 tanks, but were specifically designed for the commuter traffic from Paddington with a higher boiler pressure of 225 lbs. It was only in later years that they were more widely dispersed in the Western Region; a total of seventy engines were built.

The final class preserved is the 4500/4575 Class. They were initially built by G. J. Churchward in 1906 and the final build was completed under C. B. Collett in 1929. The 4575 Class had an enlarged water capacity compared with the 4500 Class. With smaller driving wheels than the other 2-6-2Ts they were particularly well-suited for the West Country branches – Bristol Bath Road probably had the biggest single allocation of these engines. Quite a few have been preserved and continue to give excellent service on the preserved lines.

June 1987: Standing outside of the shed at Didcot Railway Centre is the only preserved 61XX 2-6-2T, No. 6106. The shed code SLO on the front of the footplate shows that in Great Western days she was shedded at Slough, where she worked busy commuter trains in the London area.

Above: August 1988: No. 6106 is seen here at Peterborough on the Nene Valley Railway painted in unlined Great Western Brunswick green livery.

Below: Month unknown but the year is 1991: No. 6106 is standing in Washford station on the WSR with a splendid train of all 'chocolate and cream' coaches.

Above: On the same date: No. 6106 is seen here at Minehead station on the WSR preparing to couple-up to a train for Bishops Lydeard.

Below: August 1988: Pictured here at Loughborough on the Great Central Railway, 51XX class 2-6-2T No. 4141 looks resplendent in lined-out Brunswick green livery. No. 4141 was part of the post-war build programme under F. W. Hawksworth.

Above: Date unknown: Another member of the 51XX Class, No. 4140, is pictured here at Didcot Railway Centre in Great Western unlined Brunswick green livery. The small saddle tank standing alongside is No. 1340 *Trojan*; I am not sure of her origin but believe she was a Cardiff Railway locomotive.

Below: July 2006: No. 4144, a member of the 51XX Class, is pictured here inside the shed at Didcot Railway Centre. Alongside is one of the smaller 2-6-2Ts, No. 5572 – a member of the 4575 Class.

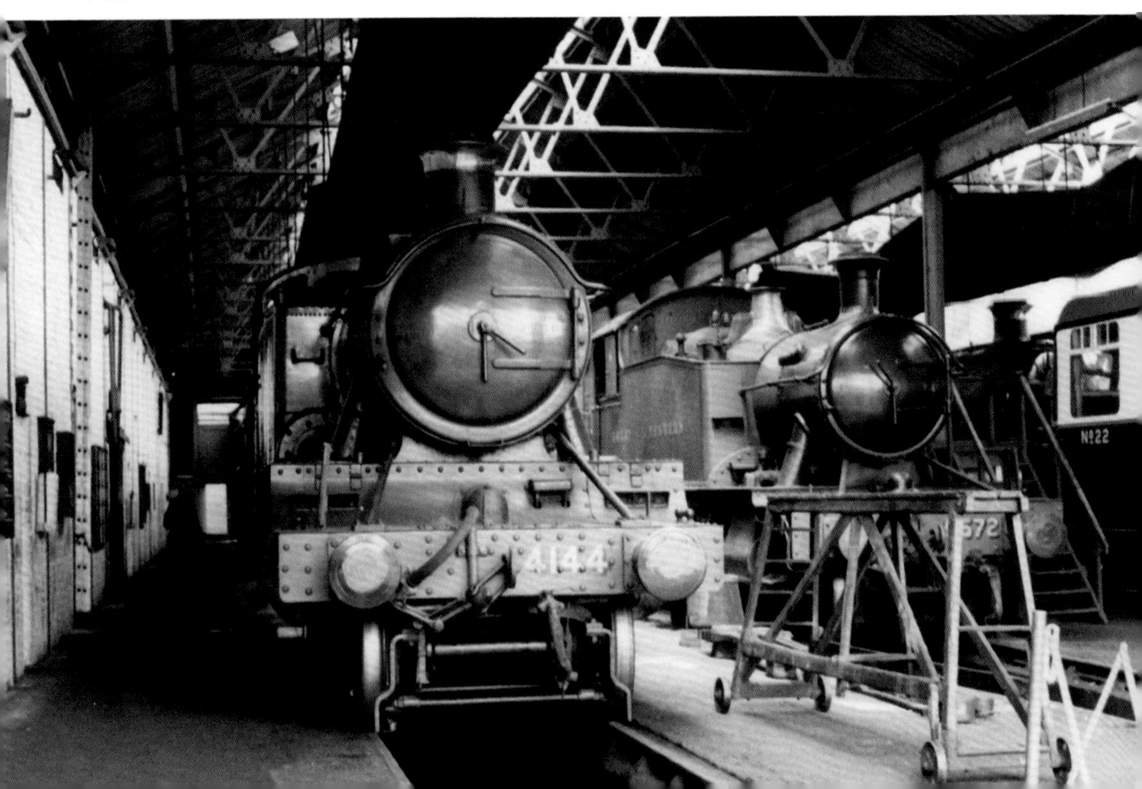

Above: August 1994: Another member of the 51XX Class, No. 4160, is well turned-out in BR mixed traffic livery. The locomotive is seen here at Minehead on the West Somerset Railway.

Below: June 1999: Same location but with a change of livery; No. 4160 is now finished in early BR livery with the wording 'BRITISH RAILWAYS' on the side tanks.

Above: September 2000: Yet another change of livery; No. 4160 is now in fully lined-out BR Brunswick green livery. She is seen here at Bishops Lydeard on the West Somerset Railway.

Below: September 2000: A final shot of No. 4160 standing outside the shed at Minehead station.

Above: June 1981: Another member of the 51XX Class, No. 5164, stands in the loco yard at Bridgnorth on the Severn Valley Railway. In the background can be seen one of her successors, a Standard Class 4 2-6-4T.

Below: June 1981: No. 5164 is pictured here at Bewdley station waiting to depart with a train for Bridgnorth.

Above: February 1983: No. 5164 is running into Highley station on the SVR. The leading coach is an F. W. Hawksworth design with sloping ends to the roof line.

Below: May 1983: A final shot of No. 5164 awaiting departure from Bridgnorth station. At the front end of the footplate she carries shed code KDR – an indication that in GW days she was shedded at Kidderminster.

Above: September 1987: Class 45XX 2-6-2T No. 4555 is seen here against the buffer stops at Paignton station on the Torbay Railway. A keen photographer, complete with tripod, prepares to take a photograph at the right-hand side of this picture.

Below: March 1991: Another member of this class, No. 4561, is pictured here at Minehead station. The loco is painted in yet another style of livery with the words 'GREAT WESTERN' on the side tanks.

Above: March 1991: Another view of No. 4561 running into Blue Anchor station en route to her final destination at Minehead.

Below: September 1995: Another location for No. 4561, standing in the yard at Bodmin on the Bodmin & Wenford Railway. With their small wheels and yellow weight coding they were ideal engines for many of the West Country branch lines.

Above: September 1995: Another shot of No. 4561 at Bodmin. Also of interest is the small saddle tank seen to the left of the picture; this was one of the two diminutive 'Port of Par' saddle tanks used on the china clay trains at Par Harbour. Both are preserved.

Below: June 1986: In the early stages of restoration, 45XX Class No. 4566 is seen here at Bridgnorth on the SVR. The early engines of this class had internal steam pipes, but I think after No. 4510 they were all fitted with external pipes.

Above: June 1990: Now fully restored and painted in lined BR Brunswick green livery, No. 4566 is on shed at Bridgnorth.

Below: June 1990: Another view of No. 4566 at Bridgnorth; she carries a shed code 83G, which according to one of my 'ABCs' refers to Templecombe, Somerset!

Above: September 1987: Pictured here on the Dart Valley Railway is 4575 Class 2-6-2T No. 4588, a variant of the original 45XX Class, built by C. B. Collett with larger capacity water tanks.

Below: September 1987: This is a typical West Country branch line scene depicted in this view at Buckfastleigh station on the Dart Valley Railway. Note there is no number painted on the front buffer beam – work in progress.

Above: September 1987: No. 4588 is ready to pull away from Buckfastleigh with a train for Totnes Riverside.

Below: September 2005: 4575 Class No. 5553 stands in fully lined-out livery adjacent to the engine shed at Minehead station.

Above: June 1988: 4575 Class No. 5572 is seen here in Great Western livery in the yard at the Great Western Centre, Didcot. The two enthusiasts at the rear of the engine are busily engrossed looking the wrong way; let's hope the loco does not move off backwards.

Below: June 1991: On one of the demonstration lines at Didcot, No. 5572 runs past the reconstructed Frome Mineral Junction signal box. This box was demolished brick by brick at Frome for rebuilding on its present site.

June 1991: No. 5572 stands at the head of a two-coach train at the demonstration line platform. The gentleman on the seat is in a typical 'spotter' mode – I wonder what's in his sandwich!

Pannier Tanks

The Pannier tanks were made up of many classes, constituting the largest type of locomotive on the Great Western. When C. B. Collett became chief mechanical engineer in 1922 he inherited a large fleet of ageing Pannier tanks – none had been built by his predecessor G. J. Churchward (1902–1921), although I believe he was responsible for converting many saddle tanks into Panniers. Collett built five classes of Panniers for varying types of work and three of these have been preserved – the 1366 Class, the 5700/8750 Class and finally the 6400 Class. His successor F. W. Hawksworth continued to build Panniers in the post-war years; he built approximately 400 of these locomotives. They were as follows – 8750 Class, 9400 Class, 1500 Class, and 1600 Class; examples of these were all preserved. It is said that the 9400 Class was built on the directive of the Great Western directors as they wished to see a more modern Pannier tank undertaking empty stock movements in Paddington station. The 1600 Class with very small driving wheels was built as a replacement for the William Dean-designed 2021 Class and finally the 1500 Class. They were a very modern design fitted with external Walschaerts valve gear, no running plate and self-cleaning smoke boxes. They had a very short wheel base and at any speed developed a very odd looking rocking gait as they moved along the track.

January 1997: This photograph was taken at a model railway exhibition at Newton Abbot. 1366 Class 0-6-0PT No. 1369 had been brought over from the Dart Valley Railway at Buckfastleigh on a low loader to form a centre piece for the exhibition.

Above: May 1998: Standing outside the shed at Bridgnorth on the Severn Valley Railway is the only preserved member of the 1500 Class, 0-6-0PT No. 1501. This class of ten locos was designed by F. W. Hawksworth principally for working the carriage sidings at Old Oak Common, although in my trainspotting days in the early 1950s I came across one at Newport Docks shed in south Wales.

Below: May 1982: A modern pannier tank in the old style. The 1600 Class pannier tanks were built as replacements by F. W. Hawksworth for the ageing 2021 Class. This photograph depicts No. 1638 on the Dart Valley Railway. Her fireman has just changed the points to give access to the coach seen in the distance.

Above: May 1982: Another view of No. 1638 standing outside the shed at Buckfastleigh; the logo 'Dart Valley' is painted in the traditional Great Western style.

Below: September 1989: Still resident on the Dart Valley Railway, No. 1638 has acquired a nameplate *Dartington*.

Above: June 1992: No. 1638 is seen at a different location, Tenterden on the Kent & East Sussex Railway. She is now minus her nameplate and finished in BR black livery.

Below: April 1989: Another class of pannier tank, 64XX Class No. 6412, is seen here arriving at Washford station on the West Somerset Railway. This class of engine was originally fitted for auto-train working, and this loco in particular became known as the *Flockton Flyer* in a TV programme of the 1970s.

Above: Month unknown but the year is 1991. No. 6412 is parked against the buffer stops at Minehead station. The buffet car to the rear was a station café popular with railway visitors.

Below: April 1992: No. 6412 is pictured here sandwiched between two other locos at Minehead station on the WSR.

Above: Date unknown. No. 6412, in black livery with a GWR 'shirt button' logo, has just arrived at Bishops Lydeard with a train from Minehead.

Below: November 1989: 0-6-0PT No. 6435 is the only other member of this class preserved, and is seen here on shed at Buckfastleigh on the Dart Valley Railway.

Above: June 1988: Another modern design of pannier tank, 94XX Class No. 9466, is seen here at the head of a short train on one of the demonstration lines at the Great Western Centre, Didcot.

Below: June 1989: Another view of No. 9466 at Didcot shed. The red route disc placed some restrictions on their availability for branch line workings, although the class numbered over 200 engines in the 32XX, 84XX and 94XX series.

Above: April 1997: No. 5764 is one of the 5700/8750 Class of pannier tank. The two classes could be easily distinguished as the 57XX Class had smaller cabs. No. 5784 is seen here at Bridgnorth on the SVW.

Below: April 1997: Another view of No. 5764, standing next to Std Class 4 2-6-4T No. 80079. The yellow route disc on the side of the pannier cab indicates a wide range of route availability.

Above: June 1981: Another view at Bridgnorth of No. 5764, this time painted in black BR livery; even the Std 4 is virtually in the same position.

Below: July 1990: A scene on the Keighley & Worth Valley Railway – 57XX Class No. 5775 at Oxenhope is painted in the fictitious livery of the 'Great Northern & Southern Railway', a livery applied for the film *The Railway Children*.

Above: July 1990: Another view of No. 5775 departing Oxenhope with a train for Keighley.

Below: June 1983: Two panniers shimmering in the sun at Bridgnorth loco yard; the leading engine finished in black BR livery is No. 7714 and the other loco in green is No. 5784.

Above: June 1983: Another view of No. 7714 at Bridgnorth carrying the shed code 85A, which denotes shedding at Worcester. In the background stands a modified Hall No. 6990, *Witherslack Hall*.

Below: March 1997: I believe this was the 150th anniversary of the Taunton–Minehead branch, now of course the West Somerset Railway. Awaiting departure from Blue Anchor station is flag-bedecked 5700 Class pannier tank No. 7760, appropriately finished in Great Western livery.

Above: September 1984: In the very early stages of restoration, the large-cabbed variant 8750 Class pannier tank No. 3650 is seen here at Didcot.

Below: September 1984: Another pannier at Didcot this time large-cab variant No. 3738 – in the background is Dukedog No. 9017 *Earl of Berkeley*.

Above: September 1984: A final view of pannier tank No. 3738 at Didcot shed.

Below: June 1988: Another view, another date. No. 3738 at Didcot shed is standing behind the diminutive 0-4-0T No. 1 *Shannon* from the Wantage Railway.

Above: July 1990: A view on the North Norfolk Railway at Holt station; pannier No. 9682 is running around her train prior to the return trip to Sheringham.

Below: July 1990: Another view of Holt station: passengers who alight at Holt station have a long walk to the small town of Holt – at the time of this visit a horse-drawn carriage was available. No. 9682 is now ready to depart for Sheringham.

July 1990: A final shot of the NNRly; this is the terminus at Sheringham, which at that time was not connected to the British Rail line that terminated on the other side of the road. These two lines have now been re-joined complete with a level crossing, making access for visiting locos much easier. No. 9682 carries the shed code 81C – the code for Southall – and pulls forward from the train which has just arrived from Holt.

Saddle Tanks and 0-4-2 Tanks

Only one class of Great Western saddle tank survived into preservation; this is the 1361 Class, designed by G. J. Churchward principally for dock shunting. A total of five were built and one survives. Several other pre-grouping saddle tanks have also been preserved; the best-known of these is 0-4-0ST No. 1338, which came into stock in 1923 from the Cardiff Railway. It spent much of its later working life shunting at Bridgwater docks, was then preserved at Uphill station (Somerset) near Weston-super-Mare and is now resident at Didcot Railway Centre.

The 0-4-2 tanks were built by C. B. Collett from 1932 onwards, and for their fairly modern construction had an antiquated appearance. Numbered in the 48XX and 58XX series, the 48s were designed for autocar working and the 58s for light goods working on branch lines; a total of ninety-five engines were built. The 48XX series were renumbered into the 14XX series in the mid/late 1940s to make room for the renumbering of the oil-burning 2800 Class, which were given numbers in the 48XX series. Although oil burning did not last for very long and the locomotives were reconverted back to coal burning, the 1400XX tanks retained their new numbers to the end.

September 1984: 1361 Class 0-6-0ST No. 1363 is pictured here at the Great Western Railway Centre, Didcot. A small class of only five locos, they were designed by G. J. Churchward in 1910 for dockyard work, and most of them were deployed at Millbay Dock, Plymouth.

Above: June 1985: No. 1363 is again seen here at Didcot as one of the exhibits in the GW 150 celebrations.

Below: Month unknown, year 1986: A final shot of 1363 at Didcot with the 'shirt button' logo on the side of the saddle tank.

Above: June 1985: Another saddle tank at Didcot for the GW 150 celebrations. The locomotive is 0-6-0ST No. 813, taken into stock from the Cardiff Railway at the time of the 1923 amalgamation.

Below: June 1989: Another shot of No. 813, this time at its original restoration site, Bewdley on the Severn Valley Railway.

Above: July 2000: 0-4-0ST No. 1340 *Trojan* pictured here at Didcot Railway Centre. I am not too sure whether I have traced her origins correctly: I believe that it is ex-Cardiff Railway, and that a sister engine No. 1338 was also preserved – I'm probably wrong.

Below: May 1990: Not all that it appears to be at first glance, this Peckett 0-4-0ST has been given the number 1144, which corresponds to a Swansea Harbour Trust locomotive, taken into stock in 1923. Pictured here on the Gwili Railway.

Above: August 1984: Pictured on the Dart Valley Railway is 14XX Class 0-4-2T No. 1420, carrying the nameplate *Bulliver*. The 14XX Class were originally numbered in the 48XX Class, but renumbering took place in the late '40s to accommodate the oil-burning 2-8-0s.

Below: August 1984: Another 14XX Class, No. 1450, at Buckfastleigh on the Dart Valley Railway. She also carried a nameplate which unfortunately I cannot read.

Above: Date unknown. 14XX Class No. 1450 is pictured here at Cranmore on the East Somerset Railway looking very smart in fully lined-out BR Brunswick green livery.

Below: March 1997: The 14XX tanks were very popular on preserved lines with Great Western histories. This picture taken at Minehead on the West Somerset Railway shows No. 1450 awaiting her next turn of duty.

Above: March 1982: 14XX Class No. 1466 at the head of a rake of preserved goods wagons at Didcot. This was not a typical duty for this class of loco as they were designed for auto-car working.

Below: March 1982: 14XX Class No. 1466 in unlined GWR livery stands in the yard at Didcot Railway Centre.

Above: September 1984: Still at Didcot, No. 1466 pulls away from the platform on one of the demonstration lines.

Below: June 1985: A final view at Didcot of No. 1466 moving away to the rear of the shed. The large boiler in the background belongs to a 72XX Class 2-8-2T.

Outside Framed 4-4-0s

In the late nineteenth and early twentieth centuries the Great Western built many O/S-framed 4-4-0s. This included the City Class: ten were built from new and a further ten were rebuilt from the Atbara Class. The most famous engine in this class is, of course, No. 3440 (*City of Truro*) – in 1904 she was timed at 102.3 mph when descending Wellington Bank on the Devon/Somerset border. This locomotive was withdrawn in 1931 for preservation but returned for a short period to revenue-earning service in 1958/9. Although no longer permitted to run on the main line, she still makes frequent appearances on preserved lines.

The last class to be built of this type were not built until 1936; they were constructed from the frames of scrapped Bulldogs and the boilers from scrapped Dukes. When originally built the first twenty were given the names of Earls, but in 1937 these titles were considered inappropriate to such old-fashioned-looking locomotives. Nameplates were removed and refitted on Castle Class locomotives, and they carried these names until the end of their working days. The 4-4-0s were numbered in 32XX series and given the unofficial name of Dukedogs by enthusiasts after the removal of their nameplates. The whole class was renumbered in the 90XX series in the late 1940s to make room for additional Collett 0-6-0s, Nos 3200–3228. One Dukedog has been preserved, No. 9017, and carries its original nameplate *Earl Berkeley*.

June 1986: The City Class O/F 4-4-0 No. 3440 is seen here at Highley station on the Severn Valley Railway. After withdrawal from service in the early '30s she was kept in static preservation until the mid-1950s when she was restored for revenue earning service on the main line, duties which she undertook for several years.

Above: June 1987: No. 3440 is on display at Didcot Railway Centre in her original GW Brunswick green livery. The loco behind carrying the headboard The Red Dragon Express is modified Hall No. 6990 *Witherslack Hall*.

Below: June 1987: No. 3440 *City of Truro* on the turntable at Didcot Railway Centre. I find it quite amazing that a locomotive of this size could have achieved a recorded speed of over 100 mph in the early 1900s.

Above: June 1987: *City of Truro* in the yard at Didcot Railway Centre; to the right of the picture is a restored LBSCR Terrier 0-6-0T.

Below: September 1984: The only other preserved class of O/S 4-4-0s is No. 3217 *Earl of Berkeley*. This photograph shows the locomotive in its original form; the nameplates were removed in 1937 and they were renumbered in the 90XX series in the late 1940s.

Above: June 1989: Given the affectionate nickname of 'Dukedogs', they were rebuilds of Bulldog frames and Duke boilers. No. 3217 is seen here in a typical branch line pose on one of the demonstration lines.

Below: September 1984: Looking at this trio of large domed Great Western engines, it may be difficult to believe they were all built in the 1930s. Just visible on the running plate of the Dukedog is the shed code WLN, the abbreviation for Wellington, Shropshire.

Narrow Gauge

At the amalgamation of the railways in 1923 the Great Western acquired two narrow-gauge lines, the Vale of Rheidol (formerly owned by the Cambrian Railway) and the Weshpool & Llanfair railways. The V of R was the last steam-operated line on British Rail until it was acquired by the Brecon Mountain Railway. At the amalgamation, the V of R had two locomotives built in 1902 and a further similar locomotive was built in 1923 by the Great Western, originally numbered No. 1203. All three locomotives are now in preservation and numbered Nos 7, 8 and 9; they are 2-6-2Ts.

The other narrow-gauge line was the Welshpool & Llanfair Railway, which originally ran from the centre of Welshpool westwards to the small market town of Llanfair Caereinion. When the Great Western closed this line the two engines were not scrapped; they are 0-6-0Ts No. 822 *The Earl* and No. 823 *The Countess*. The preservation group which reactivated the line have built a new station on the outskirts of Welshpool called Raven Square and the works and shed facilities are situated at Llanfair Caereinion. The railway also has a fairly large stock of Continental narrow-gauge locomotives and coaches which do not feature in this book.

May 1990: This photograph was taken at Devil's Bridge, the top terminus of the Vale of Rheidol Railway. The locomotive is 2-6-2T No. 9 *Prince of Wales*; she was originally numbered 1213 and built by the Great Western in 1924 to the same design as the original two V of R locos, Nos 7 and 8.

Above: July 1992: Standing in Aberyswyth station is No. 9 *Prince of Wales*, waiting to depart with a train for Devil's Bridge.

Below: July 1992: Another view of the *Prince of Wales* at Aberyswyth; the rake of narrow gauge coaches is splendidly finished in Great Western chocolate and cream livery.

Above: June 1995: Another view of No. 9, showing the large Westinghouse brake gear on the right-hand side of the engine. In the background to the right can be seen the main line standard gauge terminus of the mid-Wales line.

Below: June 1995: The plaque on the side of No. 9 is self-explanatory. The V of R was the last steam service operated by British Railways before being purchased by the Brecon Mountain Railway. As you can see I made three visits to this railway, but photographs of Nos 7 and 8 have always eluded me.

Above: June 2008: A change of railway; we are now visiting the Welshpool & Llanfair Railway, which runs from Welshpool to Llanfair Caereinion. This photograph shows 0-6-0T No. 823 *The Countess* standing by the water crane at Raven Square station on the outskirts of Welshpool.

Below: June 2008: No. 823 *The Countess*, having topped up her tanks, moves forward to the head of the next train for Llanfair Caereinion.

Above: June 2008: Looking every inch a Great Western engine, No. 823 was built by Beyer Peacock in 1902. She is ready to depart from Raven Square station with a train composed of European narrow gauge carriages.

Below: June 2008: The number plate and nameplate of No. 823 *The Countess*. The small letters GWR on the number plate show that she was taken into GW stock at the time of the amalgamation in 1923.

Above: June 1981: This photograph was taken at Didcot Railway Centre and shows the other half of the original duo of 0-6-0Ts, No. 822 *The Earl*, also built by Beyer Peacock in 1902. Sitting on a Weltrol wagon, she is obviously at Didcot for overhaul.

Left: June 2008: No. 822 *The Earl* is painted in BR black livery, looking very smart, but obviously never appeared in this livery in her working days. She is standing at the entrance to the locomotive shed at Llanfair Caereinion; her copper top chimney gives her a characteristic Great Western appearance.

Diesel Railcars

The Great Western built a small fleet of thirty-eight diesel railcars between 1934 and 1942; the class was divided into four distinctive groups. The first group were semi-streamlined and given the nickname 'flying bananas', the next group were of a more angular design, the next group of two cars were specifically designed for parcel work working in the London area. The final group of four, Nos 35–38, were single-ended units and ran in coupled pairs; as and when required, a standard corridor coach could be placed between the two units, the first DMU. Two railcars have been preserved: No. 22 at Didcot Railway Centre, which is in running order, and No. 20 at the Kent & East Sussex Railway, where I have always found it shrouded in tarpaulins. The railcars were capable of a fair turn of speed and looked very smart in their chocolate and cream livery.

March 1982: Diesel Railcar No. 22, looking resplendent in Great Western coaching stock livery, is seen here at Didcot Railway Centre.

September 1984: Another shot of No. 22 at Didcot Railway Centre. The original railcars built from 1937 onwards were much more streamlined in appearance, but the second batch completed in 1940 were given a much more angular style; in operating days I believe the valance sheeting between the two bogies was removed for ease of maintenance.

Last Build – New Build

Last build refers to the Standard 9F 2-10-0 No. 92220 *Evening Star*, the last steam locomotive built at Swindon in 1960. It was finished in traditional Great Western livery – Brunswick green with orange and black livery and a copper top chimney. After revenue-earning service it was withdrawn for preservation in 1965 – a working life of five years. It was withdrawn for preservation at the NRM and made many appearances on the main line and preserved lines; I believe it is no longer in running order but had the Great Western look about it until the end.

The new build refers to the West Somerset Railway Mogul No. 9351, built in 2004 using the boiler and frames of 51XX 2-6-2t No. 5193. No. 9351 is a smaller, lighter version of the Great Western Moguls and is ideal for preserved lines such as the West Somerset. I believe C. B. Collett actually drew up plans for a lightweight Mogul, so it has finally come to fruition.

March 1982: As all steam enthusiasts know, BR Class 9F 2-10-0 No. 92220 *Evening Star* was the last steam loco built for British Rail. Appropriately, she was built at Swindon and as far as possible was given some Great Western style embellishments e.g. a copper topped chimney and line-out Brunswick green livery. She is seen here at Didcot Railway Centre surrounded by many young enthusiasts.

Above: March 1982: No. 92220 is seen here on the demonstration line at Didcot Railway Centre at the head of a train made up of the large, recessed end-door coaching stock.

Below: May 1983: No. 92220 *Evening Star* running into Bewdley station on the Severn Valley Railway at the head of a train from Bridgnorth.

Above: May 1983: Another view of No. 92220 at Bewdley station; she is just backing off of the train which she has pulled from Bridgnorth. I believe the original No. 92220 has now been withdrawn from service, but the nameplate number and green livery have been given to another preserved 9F.

Below: September 2005: Pictured here at Minehead on the West Somerset Railway is their own-build Mogul No. 9351. Rebuilt from 5151 Class 2-6-2T No. 5193 with a Castle-type cab, she really is a very handsome-looking engine.

September 2005: At Bishops Lydeard (the other end of the West Somerset Railway), No. 9351 is running to take water from the large water tank in the background. This loco is ideally suited for the WSR, and if there are any more spare 51XX tanks I am sure another loco of this type (No. 9352) would prove very useful on many other preserved lines.